CONTENT

Words in **bold** can be found in the

WHEELS AND AXLES

Coral Martincavage

Creating Young Nonfiction Readers

EZ Readers lets children delve into nonfiction at beginning reading levels. Young readers are introduced to new concepts, facts, ideas, and vocabulary.

Tips for Reading Nonfiction with Beginning Readers

Talk about Nonfiction
Begin by explaining that nonfiction books give us information that is true. The book will be organized around a specific topic or idea, and we may learn new facts through reading.

Look at the Parts
Most nonfiction books have helpful features. Our *EZ Readers* include a Contents page, an Index, and color photographs. Share the purpose of these features with your reader.

Contents
Located at the front of a book, the Contents displays a list of the big ideas within the book and where to find them.

Index
An Index is an alphabetical list of topics and the page numbers where they are found.

Photos/Charts
A lot of information can be found by "reading" the charts and photos found within nonfiction text. Help your reader learn more about the different ways information can be displayed.

With a little help and guidance about reading nonfiction, you can feel good about introducing a young reader to the world of *EZ Readers* nonfiction books.

Copyright © 2
All rights rese
reproduced w
publisher. Prin
of America.

First Edition, 2

Author: Coral
Designer: Ed M
Editor: Morgar

Names/credits
Title: Wheels a
Description: H
Mitchell Lane P

Series: Discove
Library bound
eBook ISBN: 9

EZ readers is a

Photo credits:
Mishra from Pe
unsplash.com,
p. 8-9 Shawn H
Warrington uns
unsplash.com,
com, p. 16-17 D
rawpixel.com, p

Cars, bicycles, trucks, and roller skates! Their **wheels** are round!

RAC
333
FJB 75

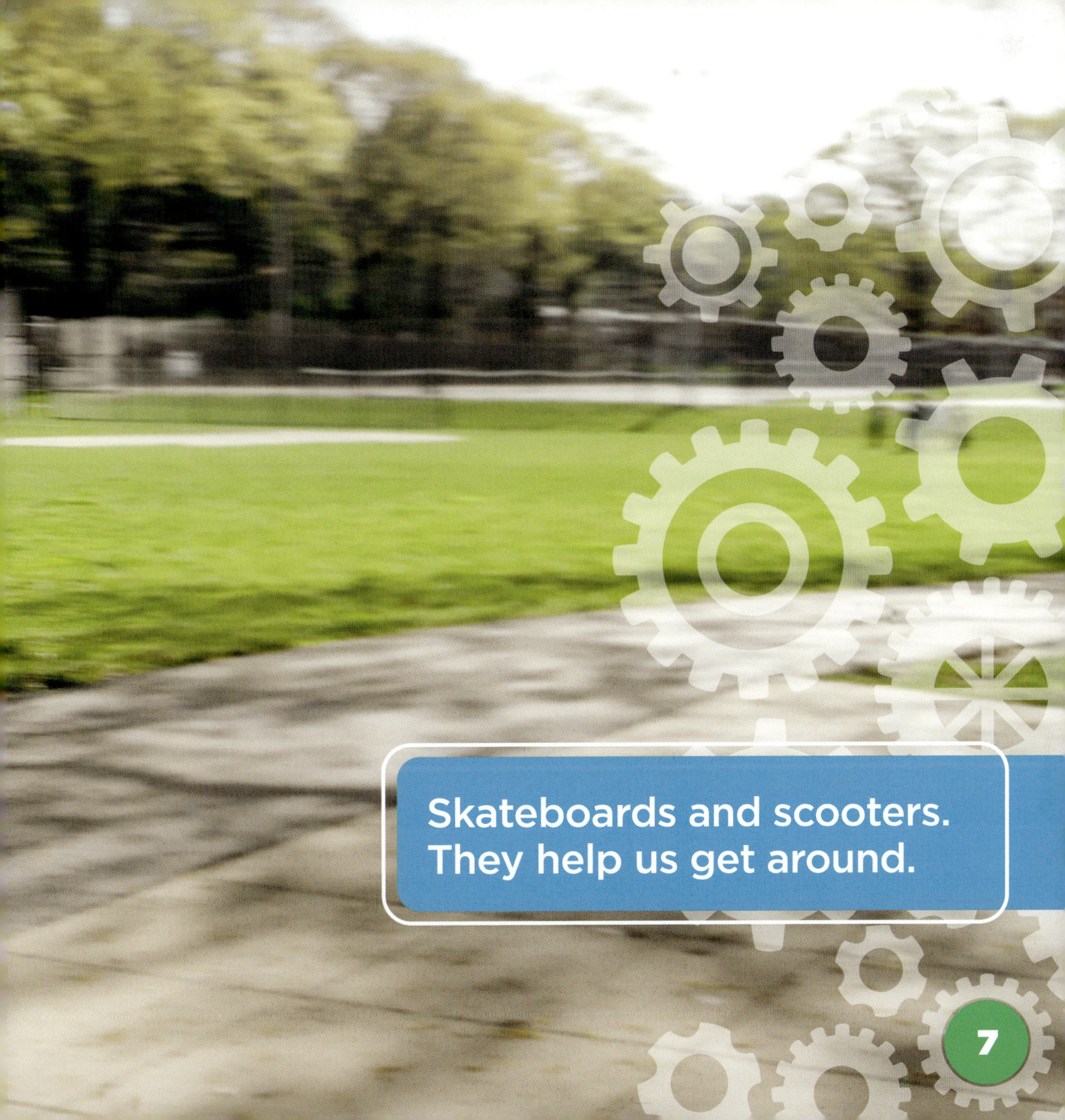

Skateboards and scooters.
They help us get around.

How do they move?
The wheels are **simple machines**.
Simple machines help to make work easier.

A wheel and **axle** is a type of lever.
They work together.
They help move people and things.

A wheel is a circle. An axle is a **rod**. The axle **connects** the wheels. It does not move.

My bike has two wheels and two axles. Can you see the wheels spin?

30 40 50 60 70
37192

The car uses wheels and axles. Axles are used for steering, driving, and braking.

Can you see me roll?
I can use a wheel and axle.

8
7
6
5

Ferris wheels turn with an axle. Wheels and axles are all around. They help us in many ways.

INTERESTING FACTS

The wheel and axle was invented in Mesopotamia about 5,500 years ago. The wheelbarrow—a simple cart with a single wheel—was invented by the ancient Greeks.

The first wheels were likely made from wood or stone, depending on the materials available.

Did you know that the wheel is one of the most important inventions of all time? The invention of the wheel changed how people and goods were **transported**. Wheels and axles also changed everyday life. Each time you use a doorknob, a pizza cutter, or a pencil sharpener, you are using wheels and axles.

GLOSSARY

axle
A rod or bar that runs through a hole in a wheel

connect
To bring or join together

rod
A straight bar, sometimes made of wood or metal

simple machine
A device that makes work easier because it uses less effort to move an object

transport
To take or carry people or goods from one place to another

wheel
A round tool that makes work easier by rotating

SOURCES

Wheels and Axles to the Rescue by Sharon Thales (Capstone Press, 1996).

Put Wheels and Axles to the Test by Sally Walker (Lerner Publications, 2011).

Wheels and Axles by Michael Dahl (Capstone Press, 2007).

FURTHER READING

Simple Machines: Wheels, Levers, and Pulleys by David Adler (Holiday House, 2016).

Wheels and Axles by Martha E.H. Rustad (Capstone, 2018).

INDEX

ABOUT THE AUTHOR

One of **Coral Martincavage's** favorite things to do is take road trips across the United States with her family. Because of the invention of the wheel and axle, road trips are possible. As a career educator, Coral Martincavage encourages children to explore the wonders of wheels and axles, find their value in everyday life, and contribute to the fascinating world of science.